AN
HOMES

written by Priscilla Hannaford
illustrated by John Dillow *and* L R Galante

Ladybird

Contents

Introduction

The Earth is home to many different kinds of animals, each with its own way of life. Many animals, such as lions, do not have a home that they come back to each night. They wander over a large area.

Other animals spend much of their time building and looking after their homes.

Nests for Homes

Most birds make nests. These are **temporary** homes, however, since the birds only use them to lay their eggs and raise their young.

Other animals, such as harvest mice, also make nests.

Harvest mouse

The harvest mouse uses grass to build a round nest. The nest is a little larger than a tennis ball and hangs between plant stalks.

Weaverbird

Some weaverbirds can tie knots using their beaks and feet, and they build their nests in this way. Hundreds of weaverbirds may build their nests in the same tree.

Different types of birds make different kinds of nests. Some birds use the same nest year after year.

Ground nest

Tree nest

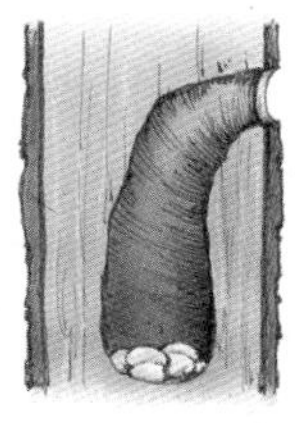

Nest in tree hole

Building on the River

Beavers are among the cleverest home builders. They build a **lodge** out of logs and mud in a pond or a lake. Sometimes they may make a lake by building a **dam** across a stream. They cut trees into logs with their sharp teeth, then float the logs into position.

Inside the lodge the beavers build a platform above the water so that they can stay dry.

Beaver lodge

Beavers enter their home by swimming through underwater tunnels. This keeps the lodge safe from enemies such as wolves or bears.

Crocodile nest

The female crocodile guards her nest to protect the eggs.

Other animals also live near the river. Some crocodiles build nests out of piles of rotting plants, which give off heat and help the eggs to hatch.

Homes in Holes

Many animals make their homes in holes in the ground called **burrows**. The burrows keep them safe, warm and dry.

Some animals spend most of their lives underground. Moles rarely come to the surface. They eat earthworms which they find in the soil. Other animals, such as badgers, come out only at night, to hunt.

In loose soil, moles can dig one metre every three minutes.

Fox

Foxes come out of their holes at night to hunt birds, rabbits, mice, **poultry** and frogs.

Earthworms

Earthworms make holes in the ground by swallowing the soil in front of them.

Rabbits

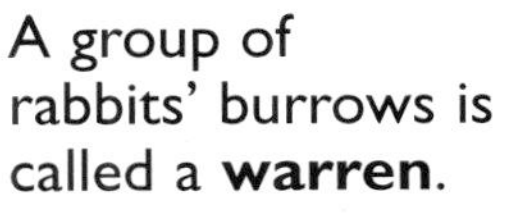

A group of rabbits' burrows is called a **warren**.

Insect Homes

Wasps, bees, ants and termites all build nests to protect their young.

Most bees like to live in large groups. Each group has a queen bee, who spends her time laying eggs. Worker bees are female, and male bees are called drones.

Honeycomb

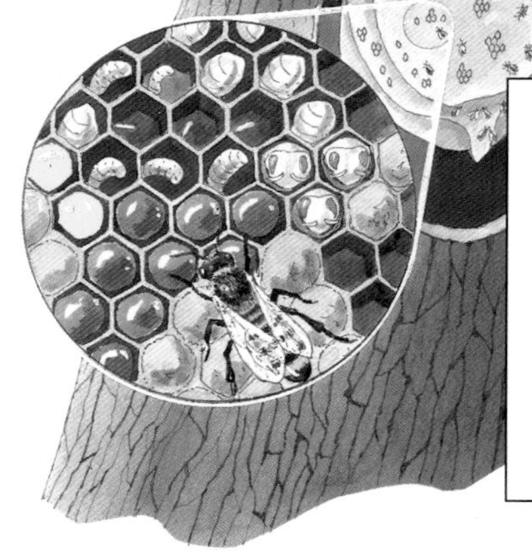

Bees make honeycombs out of wax cells. The bees store honey and pollen in some cells, and the queen lays her eggs in others.

Most of an ants' nest is underground. The pile of earth we see is what they have dug out of the ground!

Each nest is a maze of tunnels with rooms leading from them. Some rooms are used for storing eggs.

Ants and termites build the biggest and most complex insect homes. A large termite nest may have more than five million termites in it.

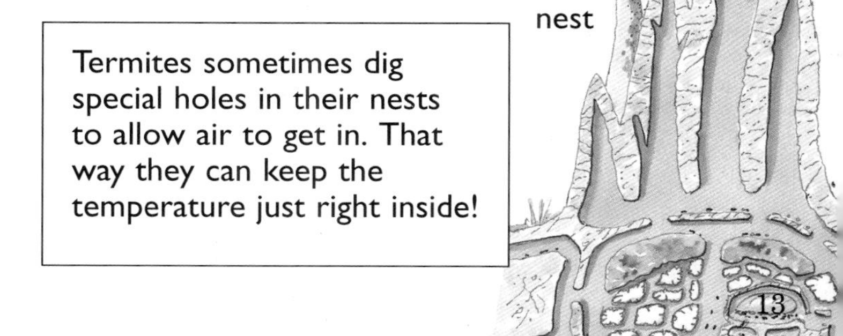

Termites sometimes dig special holes in their nests to allow air to get in. That way they can keep the temperature just right inside!

A Home in the Rainforest

Rainforest trees are very tall and grow close together. Some animals spend their lives high up in the trees. Others, like the jaguar, live on the forest floor. The trees are so tall that jaguars may never see the animals living near the tops of the trees.

Gorillas build sleeping platforms out of twigs and branches on the ground or in low branches of trees.

Gorillas

The toucan lives near the tops of the trees, its colours helping it to blend in with the flowers that grow there.

The hummingbird is the smallest bird in the world. Its nest is held together with spiders' webs.

Orangutans live in the middle layer of the rainforest, and rarely come down to the forest floor.

Flying squirrels sleep in hollow trees. They hunt at night, gliding from tree to tree.

A Lion's Territorial Home

Lions live in large family groups called **prides**. Each pride has its own **territory**. Male lions protect their territory from other lions. Their warning roar can be heard eight kilometres away.

A female lion is called a **lioness**. The lionesses do most of the hunting, usually at night.

Each pride of lions has six to thirty members. There are usually one or two adult males in each pride.

A pride of lions

Male lion

Male lions are fairly lazy and spend most of the day resting in the shade.

It is easy to tell the difference between a lion and a lioness, since only the male lions have manes. Lionesses are smaller and a lighter colour.

Lioness

Female lions are very good mothers. Baby lions are called **cubs**.

Polar Homes

The North and South Poles are the coldest places on Earth. All the **polar** animals that live there have special **features** to keep out the cold.

Polar bears, beluga and narwhal whales and walruses all live near the North Pole. Penguins live near the South Pole.

Whales have a very thick layer of fat, called **blubber**, which keeps them warm.

Beluga whale

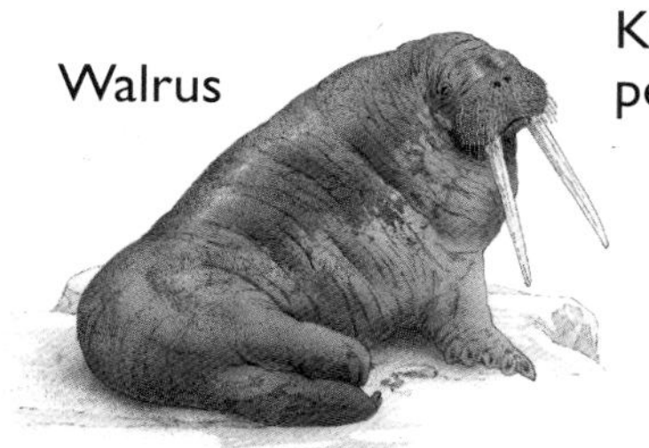

King penguins

Walruses sometimes sleep in the water, hooking onto the ice with their tusks.

When a female king penguin lays an egg, she rests it on her feet. The male and female take it in turns to hold the egg until it hatches.

A female polar bear gives birth to her babies in caves made out of snow.

Polar bears

Plants as Homes

Plants often provide food and protection for animals. Many insects live on plants.

Animals that live on plants and trees are often the same colour as the plant. Their colouring makes them difficult to see and protects them from their enemies.

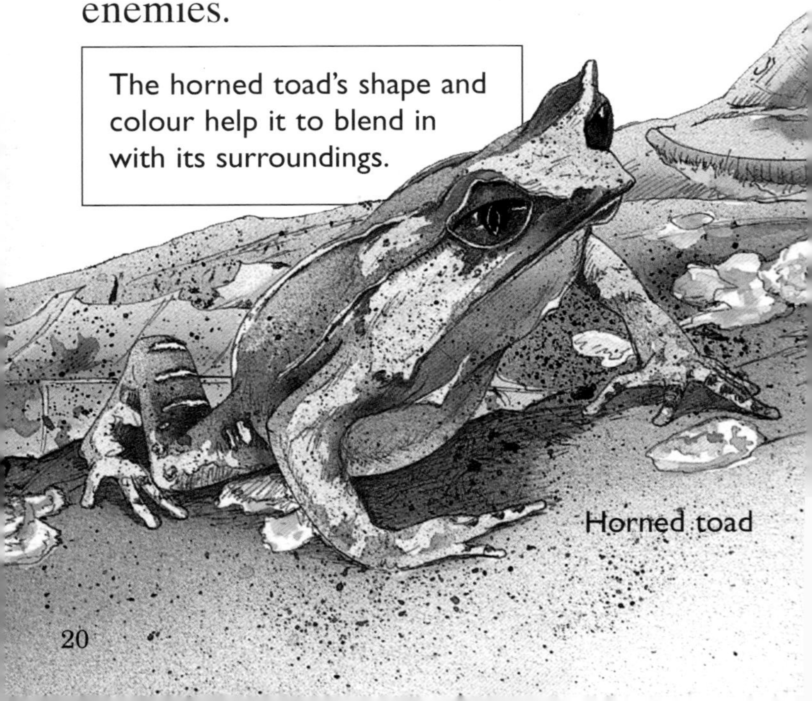

The horned toad's shape and colour help it to blend in with its surroundings.

Horned toad

Larger animals also use plants as homes. Chimpanzees build nests in trees to sleep in, to protect themselves from lions and other animals. They make the nests by twisting small branches and twigs together.

Chimpanzees are very good at climbing trees, but they spend more of their time on the ground.

A family of chimpanzees

The River as a Home

Otters spend part of the time on land and part of the time in the river. They often make their homes under the roots of trees, along the riverbank.

An otter's home is called a **holt**. The entrance may be underwater, so that the otter can enter and leave the holt without being seen.

Otter

When an otter catches a big fish, it usually brings it to the bank before eating it.

Kingfisher

The kingfisher makes its nest in a hole in the riverbank.

The female stickleback lays her eggs in a nest on the riverbed. Then the male guards the nest until the eggs hatch.

Stickleback

The stickleback makes his nest from pieces of water plants and other materials.

Secondhand Homes

Some animals use homes that belonged to other animals. Foxes may use holes dug by badgers, for example. The hermit crab doesn't have a shell of its own. It lives in an empty shell that it finds lying on the sea floor.

If the shell gets too small, the hermit crab moves to a bigger shell.

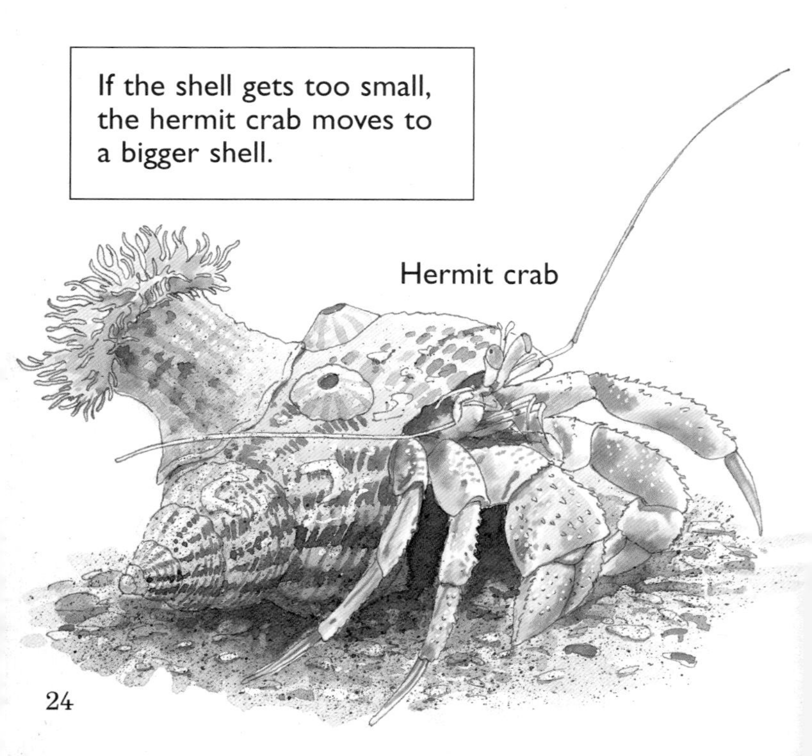

The mother sparrow feeds the big baby cuckoo as if it were her own chick.

Baby cuckoo in sparrow's nest

The cuckoo always lays her egg in another bird's nest. When the baby cuckoo hatches it pushes the other eggs out of the nest.

Clown fish

The clown fish makes its home with the poisonous sea anemone, where it is safe.

Hibernation

Some animals go into a special long sleep, called **hibernation**. They crawl into a cave or nest and go to sleep in the late autumn. When springtime comes, they wake up. The dormouse is one of these animals.

Tortoise

Tortoises go into a deep sleep in cold weather. They sleep in the mud or earth, or under plants, until the weather becomes warmer.

When animals hibernate their body temperature drops until it is almost as cold as the air around them. Their heartbeats become faint and their breathing slows down.

Dormouse

The dormouse curls into a tight ball to hibernate.

Squirrels sleep for most of the winter, but wake up and move about on warmer days.

Squirrel

Pet Homes

Many people have animals as pets. Some pets can live in our homes, like cats and dogs. Other bigger animals, like horses, need their own homes.

Taking care of a pet is hard work. You have to make sure that it has enough food to eat and water to drink. Your pet needs somewhere dry and warm to sleep and you have to give it plenty of exercise.

This pet is well looked after. It can sleep soundly in its dog basket.

If you have a horse, you must clean out its stable every day.

Stable

Hamster's cage

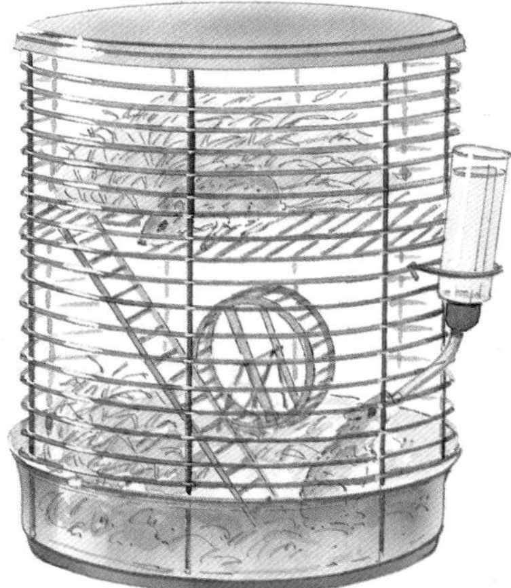

You must make sure that your hamster always has water to drink.

Glossary

Blubber A special layer of fat that animals that live in cold places have to help them to keep warm.

Burrow A hole in the ground made by an animal, to live in.

Cub A name given to some animal babies. Baby lions are called *cubs*.

Dam A barrier built across a river or stream, to stop the water from flowing.

Feature Something that is special about a particular thing.

Hibernation A special kind of deep sleep that lasts all winter.

Holt An otter's home.

Lioness A female lion.

Lodge A beaver's home.

Polar Something relating to either the North or South Pole.

Poultry Farm birds like chickens, ducks, turkeys and geese.

Pride A group of lions.

Temporary Something that lasts only a short time.

Territory An area of land that is defended by a bird or animal.

Warren A group of rabbit burrows.